I0503216

Drought is Not
A Four-Letter Word

A Citizens Guide to New Water

by
Mohammed A. Hasan
M.S., P.E., F.ASCE, PWLF

with research and writing assistance
by Tim Pompey

Published by Mohammed A. Hasan
Copyright © 2015 by Mohammed A. Hasan
All Rights Reserved

ISBN: 13-978-1516906796
ISBN: 10-1516906799

Cover Design:
Nancy DeLucrezia

This book is dedicated to those individuals
who will take these proposed water concepts
and tirelessly move them forward.

Much thanks to:

Mark Capron, P.E., and my other partners
in PODenergy, Inc.
and
Tim Pompey for his writing and research assistance
and John Weigle

I

Drought is not a four-letter word

Drought is here
Drought was here
Drought will be here

So...

☞ Before they tell you to drink toilet water . . .
☞ Before they tell you to remove your lawn . . .
☞ Before they tell you to uproot almond trees . . .
☞ Before they scare you with higher costs for water . . .
☞ Before they penalize you for saving water by jacking up prices . . .
☞ Before your water is rationed . . .

Water Engineers, Scientists, and Managers

✔ Must find new water by inventing and innovating . . .
✔ Must re-educate the public about where water comes from . . .
✔ Must involve everyone in the search for water . . .

✔ Must tell people what can be done to overcome the severity of drought . . .

✔ Should facilitate engineering alternatives with full support to replace outdated technologies . . .

✔ Should never back away from trying new things to solve old issues . . .

Danger, Danger

☺ If action is not taken now to find a sustainable solution, the problem will continue . . .

☺ We will hear this same saga every few years . . .

☺ Bigger problems necessitate very large projects . . .

☺ Very large projects take decades and billions of dollars and most important, time . . .

☺ Doing nothing or only conserving is not going to be forgiven by coming generations . . .

Change is essential

Change is happening all around us, including population explosions and the increased effects of climate change. The result is less availability of water for everyone to use.

Similarly, water availability from the rich bay Delta area in Northern California faces challenges of not just climate change but also political opposition and infighting.

While the whole world has changed in the last fifty years, while severe climate change is becoming more scientifically evident, and while severe drought is taking hold of California in historical fashion, our water supply has not changed. Human life, in constant need of water, is being put at risk. Only by changing our thinking and our approaches to new water technologies can we benefit the thirsty people of California.

II.

A Little History About Drought
in California

It's possible that those of us who have lived in California since the mid 20th century have been fooled by an unfounded assumption. That assumption is quite simple: California is wet in the winter and dry in the summer. In Southern California, we've taken it for granted that rain and snow will appear sometime between November and April of each year. It will come in varying amounts, but it will come nonetheless, refill our aquifers, rivers, and snowpacks, and allow us to continue using water with only minimal forethought.

Of course, some years will be drier, some years wetter, but all in all, the water we've used for lawns, showers, sprinkler systems, and agriculture will cycle through fairly predictably, and we'll continue on a daily basis with what we consider "normal" water usage.

But the scenario may be changing. What if one day we wake up, turn on the spigot, and nothing comes out? Really, you ask? As you will learn, it's not so far fetched, and in some towns in California it's actually happening now. Rivers, lakes, and wells have dried up, and residents are desperately seeking to replenish their water supplies.

A little examination of drought in California shows that what we have assumed about water cycles may be incorrect and that we've built our public water needs and infrastructure around a cycle of rain and snow that varies drastically from decade to decade and century to century.

In addition, since the Dust Bowl era of the 1930s, California's population has swollen dramatically in areas that are notorious for desert terrain and drought. Los Angeles, for example, which is built on the edge of the desert, was listed in 2012 by the U.S. Census Bureau as the most densely populated urbanized area in the U.S.

To date, California has suffered through its fourth year of drought. With almost 40 million people living in this state and a 50 billion-dollar agricultural industry to support, it's possible that we may have bought into this illusion about California's water cycles for too long and pushed our cities and counties into a perilous position. Too many residents. Not enough water.

—

Some are calling this California cycle of drought one of the worst on record, but studies of such things as ancient tree rings tell us that there have been more severe droughts in California. In fact, the rainy cycles we've taken for granted may not be normal at all. They may be variants in a series of dry cycles that can last for decades or centuries. As an example, the last strong El Niño in the state was more than 15 years ago. Since then, it's been more dry than wet.

According to a recent New York Times article written by Henry Fountain,[1] there have been previous "megadroughts" that have lingered in the state for more than two centuries.

"Tree ring analysis has revealed many historic droughts," Fountain wrote, "including one throughout much of the Southwest around 200 A.D. that lasted for five decades." Other studies have revealed that one drought in the 9th century lasted about 200 years, and another in the 13th century lasted about a century and a half.

The Dust Bowl itself should give warning enough that such droughts are possible and can be devastating. According to Erin Carlyle of Forbes,[2] the Dust Bowl lasted for eight years and at its peak in 1934 covered 77% of the U.S. Oddly enough, one of those areas that remained farmable was California.

This time around, however, we may not be so lucky and the cracks in our water delivery infrastructure are starting to show. For instance, in East Porterville,[3] where

---

[1] Henry Fountain, "In California, a Wet Era May Be Ending," *New York Times*, 13 April 2015, Science Section.

[2] Erin Carlyle, "Beyond California: The Worst Droughts in American History," *Forbes*, 13 May 2015.

[3] Jessica Glenza, "The California town with no water: even an 'angel' can't stop the wells going dry," *The Guardian*, 20 April 2015.

nearly a thousand wells that once delivered water to mostly rural farm folks have dried up, forcing the town's residents to take drastic measures such as hauling in water to their area.

All of this is to say that the illusion of wet and dry seasons is starting to break apart and Californians are starting to panic. The State is pushing its citizens to conserve. Cities are tightening watering restrictions. Regional political dog fights are breaking out over water rights.[4]

But what if the solution to all of these problems is not rationing or other drastic measures, but rather acknowledging, working in partnership, and living with the history of our own geography?

What if we take a broader approach and plan for a future in which water is something to be studied, measured, conserved, but ultimately expanded through science and recent advances in technology?

What if with a little forethought, we could find ways to ensure that everyone has plenty of water? Instead of depending almost exclusively on Mother Nature, what if we can use current technologies to advance our cause without doing major harm to the environment? Without robbing one water source to use somewhere else? Thinking

---

[4] Nick James, "Senior Water Rights Holders Take California Water Board To Court Over Drought Restrictions," *CBS News Sacramento*, 19 June 2015. <http://sacramento.cbslocal.com/2015/06/19/senior-water-rights-holders-take-california-water-board-to-court-over-drought-restrictions>

broadly, this would encourage more cooperation and less desperation. It would ease the pressure on our current water systems and give us a sense that the problem is manageable.

Instead of spreading panic, science could help us to develop solutions and engage in cooperative sharing. It's not easy. It's not cheap. But it is doable. The resources are there. The science is there and has already been put to use in other drought stricken countries like Australia.

With this book, we choose to engage in fruitful conversation, and we hope that this will help demonstrate how these problems can be solved, not just among scientists, but among people from all different backgrounds who share a similar concern about increasing water supplies.

III.

Human Density vs. Desert Terrain: The State in which We Live

The word "density" can be used in two ways, and for our purposes, both definitions suit our purposes quite well.[5]

1. The quality or state of being dense.
2. The quantity of something per unit volume, unit area, or unit length.

Let's see how both of these definitions apply to water management in California.

1.

In California, density applies to both the state's populations and its use of land, i.e., agriculture, and in both cases, it directly affects the use of water.

---

[5] The Merriam-Webster Collegiate Dictionary, Eleventh Edition, s.v. "Density."

The deserts of the Southwest for centuries sustained human life, both tribal and communal. Then came the era of the population boom, particularly in areas like Los Angeles, Phoenix, and Las Vegas, all in different states. Suddenly water came at a premium. When government eyes looked around for regional water sources, they settled on the Colorado River, a wild, winding source of water that was eventually tamed through dams and lake reservoirs. The river now serves as the main source of water for several western states, but given the increase in urban density, science is demonstrating there simply isn't enough Colorado River water to supply urban areas like Los Angeles, Las Vegas, Phoenix, Tucson, San Diego, and many others.

According to the journal ScienceDaily, the American west now has a serious "drinking" problem. Scientists estimate that to live sustainably, "we should use no more than 40 percent of the water from the Colorado basin. As of now, we use 76 percent."[6]

Much to the consternation of farmers in the center of the state,[7] in the early 1900s, Los Angeles began to build an

---

[6] "Water usage far exceeds sustainability level in the desert southwest, US," *ScienceDaily*, as sourced from Arizona State University, College of Liberal Arts and Sciences, 10 May 2012. <http://www.sciencedaily.com/releases/2012/05/120510224444.htm>

[7] Thor Benson, "Can America's Desert Cities Adapt Before They Dry Out And Die?" *Fast Company*, 9 December 2014

aqueduct from the mountains of Northern California directly to the city, Despite their protests and even some violence, the aqueduct was built and today the Owens Lake, once a spreading basin of water in the central valley of California, is completely dry.

However, as the drought has spread, the snowpack on which the aqueduct relies to funnel water south has all but disappeared. Last year in particular was one of the driest on record. The most recent snowpack measurement last spring couldn't be done because there was no snow to measure.[8]

2.

Density in its formal sense also applies to agriculture in the state of California. Agriculture in the southwest is estimated to use 77 percent of water allocated for human use.[9]

In California that estimate is as high as 80 percent. The problem is that while generating nearly 50 billion dollars worth of produce, agriculture's share of the state's economy is only about 2 percent.[10]

---

[8] Jon Erdman, California's Snowpack at Record Early-April Low; Sierra Snow Survey Finds Bare Ground," *The Weather Channel*, 10 April 2015.

[9] Ibid, ScienceDaily.

[10] Jeff Guo, "Agriculture is 80 percent of water use in California. Why aren't farmers being forced to cut back?" *The Washington Post*, 3 April 2015, GovBeat.

This is not to downplay the importance of agriculture, since California serves as one of the world's major breadbaskets, but to service this breadbasket requires large amounts of water sucked from terrain that is literally drying up. Some crops are more thirsty than others. A single almond, for instance, takes about a gallon of water. A single walnut takes about five gallons of water.

Add to this the fact that at the same time as the drought has kicked into full force and California farmers are fighting for water rights (the politics of which we will discuss later), the state's export of nuts to places like China is skyrocketing.[11]

Human density. Agricultural density. They have mushroomed in California and made it difficult to find enough water to go around. The result is similar to a school yard full of bullies seeking to knock each other around and ensure their own water privileges. It's not a sustainable model. Someone wins, someone loses, and water itself becomes a source of nonproductive wrangling.

While some rationing and conservation might be appropriate during this time of drought, the method for servicing both cities and agriculture is based on uncertain assumptions about weather and land use, particularly in a state that is notorious for dry terrain.

We suggest that perhaps there is a more efficient way to operate, a way to produce and distribute water that

---

[11] Julia Lurie, "California's Almonds Suck as Much Water Annually as Los Angeles Uses in Three Years," *Mother Jones*, 12 January 2015, Environment.

makes more sense for the populations and the produce grown within this state.

Rather than robbing from Peter to pay Paul, or standing guard over what water rights may be legal in the state, or even reducing the amount of water distributed to farmers or cities, there may be a better way to operate that allows more sharing and more water to be available throughout the state.

This then becomes our major goal: To see that water is treated as a reasonably serviceable product, much the same as milk or gas, or some other commodity that relies on cost efficiency and fair pricing and the ability to deliver over a wide area of distribution.

There is technology to do that. Ask Australia. The question for us is whether or not we can cooperate within the state, even from state to state, before the damage falls beyond our control.

We wish to avoid the uncharted territory that results when the severity of drought catches us by surprise. Imagine these situations:

Water agencies only allow us to use water in the morning and in the evening, meat and other food prices triple, restaurants, stores, coffee shops and bars reduce their hours. Property values go down while the cost of water skyrockets.

These and many other scenarios can likely happen if we're caught off guard and do not plan today.

IV.

A Not So Simple Question: Where Do We Get Our Water?

California has a long history of water development, especially in the Southern California area. But where exactly does this water come from?

Of course, most local water resources include surface water from lakes and rivers as well as water from underground springs and aquifers, commonly known as groundwater. Indeed the history of water in California starts with the digging of wells. Most areas depend on some type of groundwater basin to provide water for their municipalities and farmlands.

However, as urban areas in California began to grow rapidly, especially after the gold rush in 1849, more water was needed to sustain housing, farming, and industrial development.

As early as 1873, President Ulysses S. Grant commissioned a study by the U.S. Army Corps of Engineers to survey the Central Valley's irrigation needs. The report recommended "the systematic development of the Sierra watersheds."[12]

After decades of planning and political wrangling, along with some federal intervention during the depression in the 1930s, construction began in 1935 on what we know today as the Central Valley Project, which currently supplies water to the San Joaquin Valley and the Bay area.

Southern California had its own unique set of problems, mainly a booming population living in a desert terrain. It worked on solving its water problems with two large construction projects.

The first was the building of the Los Angeles Aqueduct, which carried water from the Sierra Nevada Mountains two hundred miles south to Los Angeles.

In 1904, the Los Angeles board of water authorized William Mulholland and several other engineers to look for additional water sources. One of the areas identified was the Owens River, which flowed out of the eastern side of the Sierra Nevada Mountains and formed Owens Lake. A bond for $1.5 million was issued and passed by voters from Los Angeles in 1905.[13] Construction started on the Los Angeles Aqueduct in 1908 and was finished in 1913. When completed, it was the world's longest aqueduct, at 233

---

[12] "History of Water Development and the State Water Project," California Department of Water Resources, History of the California State Water Project. <www.water.ca.gov/swp/history.cfm>

[13] "Los Angeles Aqueduct," *History*. <http://www.history.com/topics/los-angeles-aqueduct>

miles (375 kilometers), and the largest single water project in the world.

In 1925, the Los Angeles Department of Water and Power (DWP) was established and the city voted to approve a $2 million bond issue to begin development of the Colorado River Aqueduct.[14] In 1928 the DWP agreed to a partnership with its regional cities to form a state district that is now known as the Metropolitan Water District of Southern California (MWD). The result was the completion of Hoover Dam in 1935 and the Colorado River Aqueduct in 1941. This partnership now includes 14 cities, 12 municipal water districts, and a county water authority.

In 1963, the state of California began construction on what is known as The Governor Edmund G. Brown California Aqueduct system of canals, tunnels, and pipelines that carry water collected from the Sierra Nevada Mountains and valleys of Northern and Central California to Southern California. The aqueduct took ten years to build and was officially opened in 1973.[15]

The Department of Water Resources (DWR) operates and maintains the California Aqueduct, including one pumped-storage hydroelectric plant, Gianelli Power Plant. Gianelli is located at the base of San Luis Dam, which

---

[14] "The Colorado River: A Regional Solution," City of Los Angeles Department of Water and Power. <http://wsoweb.ladwp.com/Aqueduct/historyoflaa/col oradoriver.htm>

[15] "California State Water Project Milestones," California Department of Water Resources. <http://www.water.ca.gov/swp/milestones.cfm>

forms the San Luis Reservoir.

The Castaic Power Plant, owned and operated by the Los Angeles Department of Water and Power, is located on the northern end of Castaic Lake, while Castaic Dam is located at the southern end.

—

Today, most cities in Southern California use a combination of sources to provide water to their customers. Various cities around Ventura County have access to different water resources. For cost purposes, Oxnard blends together water from the United Water District, from well water, and from the Calleguas Water District.

For example, the city of Oxnard gets water from three different sources: underground wells, from the United Water Conservation District of Santa Paula, and water from a mix of the California Aqueduct and the Colorado River Aqueduct. This is done via the Santa Susanna tunnel which flows into Ventura County.

Other Ventura County cities use different sources. Ventura gets its water from Lake Casitas, from the Ventura River, and from underground wells. Ojai depends on water from Lake Casitas and from the private water company, Golden State Water. Santa Paula and Fillmore both depend on large underground aquifers.

V.

The True Cost of Water:
What Are We Paying For?

Understanding the true cost of water in California is a very tricky issue, especially as cities and regions try to deal with the state's serious drought issues. In short, there is no easy answer to this question because it depends on where you live and what water sources your city or county are using. Your water purveyor[16] is the one that supplies water to you.

To complicate matters, as cities try to lower their water usage to deal with drought conditions, the California courts have waded into the matter and made clear that, at least for now, cities are limited in what they can charge. A recent ruling by a state appeals court said that a tiered water rate structure implemented by the city of San Juan Capistrano, which implemented a tiered system charging

---

[16] Your water distributor or supplier

different rates, depending on how much water a customer used, was unconstitutional.[17]

According to the court, the structure violated California State Proposition 218, which prohibits state government agencies from "charging more for a service than it costs to provide it."[18]

What still remains possible, as is the case in the city of Santa Cruz, is that users can be penalized for using too much water. So, as the drought continues, cities are wrestling with answers to these questions: How much water usage can we allow and what methods can we use to encourage conservation? Over the next several years, as the State government and courts deal with this question, your water rates could vary considerably.

As for general cost determination, we can use Oxnard as a local example of a typical rate structure. Oxnard bases its rate structures on two water sources. The city buys water from the United Water Conservation District and from the Calleguas Water District.

United is located in Santa Paula. It diverts water from the Santa Clara River during heavy rainfalls that is then percolated into the groundwater basin. Wells pump this water to Oxnard.

---

[17] Matt Stevens of the Los Angeles Times and Bay Area News Group staff writers Paul Rogers, Jeremy Thomas and Denis Cuff, "California drought: Court rules tiered water rates violate state constitution," *San Jose Mercury News*, 20 April 2015, Science and Engineering.

[18] Ibid.

Calleguas receives its water from the Metropolitan Water District. Its water is a combination of water from the California State Water Project and is delivered by the California Aqueduct and also the Colorado River Aqueduct.

The price for Oxnard's water is based on a usage term known as "acre feet." Here is the general terminology for determining water rates:

- 1 acre foot = 326,000 gallons.[19]
- Customers are billed per 100 cubic feet of usage
- 1 cubic foot = 7.48 gallons.
- 100 cubic feet = 748 gallons.[20]

Oxnard uses a tiered structure according to how much is used per month. So the more you use, at a certain amount of usage, or a tier, the higher the cost.

While this is a general example, other cities may use different means to determine their water rates. To learn how water is billed in your city, you should browse carefully through your utility bill, or call your city's utility department.

Some cities have water supplied through private companies. Ojai, for example, is tied to Golden State Water

---

[19] A foot of water on a flat surface that is an acre in size (AF).

[20] This is how you are normally billed (HCF).

Company. A private company's water rates must go through the state's Public Utility Commission.

For each individual city, water rates are determined by the city council, but the recent drought has left cities in California struggling to try to figure out what it means to charge for water without exceeding their costs.[21] For example, Santa Barbara has resurrected its desalination plant.[22]

When it comes to farming, a different set of agencies and rate structures apply. Most farmers, especially in California's Central Valley, get at least part of their water from the California State Water Project and the federal Central Valley Project. A farmer may have what is known as "senior water rights" on water sources such as wells that exist on the property.

It's important to note that water in California is a "property right," meaning farmers have a right to use water on their property, but they do not own the water itself.[23] As a result, they are subject to a complicated set of procedures for managing their water use.

---

[21] Editorial Board, "What's next on California's water rates?" *Los Angeles Times*, 22 April 2015.

[22] Amanda Covarrubias, "Santa Barbara working to reactivate mothballed desalination plant," *Los Angeles Times*, 3 March 2015, California.

[23] "The Water Rights Process," California Environmental Protection Agency, State Water Resources Control Board. www.waterboards.ca.gov/waterrights/board_info/water _rights_process.shtml

According to the State Water Resources Control Board: "A water right is a legal entitlement authorizing water to be diverted from a specified source and put to beneficial, nonwasteful use. Water rights are property rights, but their holders do not own the water itself.

The exercise of some water rights requires a permit or license from the State Water Resources Control Board, whose objective is to ensure that the State's water resources are put to the best possible use."[24]

The State of California has in place several different types of water permit processes. Based on state law and water availability on their property, a farmer can determine their usage and cost based on what water is available to them in their area. This means that water costs as a part of their business structure may vary, and depending on a farmer's contract with a water supplier, their rates could be cheaper than the average consumer. Still, every farmer across the state has to submit a plan to the county and/or state regarding their forecast for water sustainability

Even with certain water rights in place, farmers may not be able to draw as much water as they wish because there may be state and local restrictions on how much they can use on an annual basis.

In light of California's recent drought, this has been controversial because of the water restrictions which Governor Brown implemented statewide in April 2015.[25]

---

[24] Ibid.

[25] Janell Thomas, "California implements water restrictions as drought continues," *Farm Futures*, 2 April

These restrictions have caused a stir among individual citizens, cities, counties, and farmers because of the difficulties imposed on them by state law and state water policy. As we see in the court case with San Juan Capistrano and in proposition 218, interpretation of water policy in California is still very much a matter of debate.

Based on those restrictions and the possibility of diminishing water resources, some farmers are saying that they are going to reduce the amount of land they farm.

"I'm going to fallow two acres of my land immediately," said Geoffrey Galloway, a citrus farmer from Porterville. "Depending on how the season goes, we may let another four go."[26]

While it's true that some farmers may have advantages in accessing and purchasing water, depending on their water rights and water availability on their property, but they all face a common problem. Their sources are subject to being interruptible. So, in a serious drought such as California is currently experiencing, many farmers are now

2015.
<http://farmfutures.com/story-california-implements-water-restrictions-drought-continues-0-125936>

[26] Adam Nagourney, "As California Drought Enters 4th Year, Conservation Efforts and Worries Increase," *The New York Times*, 17 March 2015.

worried about whether or not their water supply will be drastically reduced or completely shut off.[27]

It's an economic, political, and social dilemma that has left state regulators and politicians fighting over decreased water options. At issue, however, is the survival of the state's water resources as well as the survival of all of us who depend on water.

The questions should be asked: Is conservation our primary answer, or are there other options? Maybe new technologies and innovative ideas that have been overlooked? Could there be a way to consider those options without all the infighting? Maybe it's the fighting itself that has muddied the waters.

We contend there are more solutions to water resources than have previously been proposed, considered, and implemented. The real problem itself is, in order to solve our water issues, can we listen to each other?

---

[27] Nick Janes, "California Farmers Worry Senior Water Rights Cuts In Drought Could Be Devastating," *CBS News Sacramento*, 12 June 2015. <http://sacramento.cbslocal.com/2015/06/12/california-farmers-worry-senior-water-rights-cuts-in-drought-could-be-devastating>

VI.

Who Runs the Show:
Water Agencies, State Government, and Public Ownership

As we noted earlier, water rights in California are actually "property rights." Those who have water on their property own the rights to the water based on their property rights rather than water ownership. For this reason, water has become a managed resource via various federal, state, and county agencies. It's like a pie in which many different hands are involved in regulating various size pieces of that pie. It is helpful, then, to know who's involved and why water management in California is so complex and often convoluted.[28]

First, we must understand that the reason water is such a different type of resource from other natural products like oil is because:

---

[28] For a more detailed history of water management in California, read Ellen Hanak, et al., *Managing California's Water: From Conflict to Reconciliation*, (San Francisco: Public Policy Institute of California, 2011).

- It's widespread
- It's vital for survival and there's a public interest in managing it, and
- It's replenishable.

Briefly, we're going to mention the various agencies in the state that are directly involved in water management. We do this not to confuse you but rather to give you a sense of the difficulties we as citizens face when it comes to knowing how our water is doled out, sold, and regulated.

Federal

It comes as no surprise that first and foremost when it comes to water management, we must look at the Federal government, which is intimately involved in setting rules regarding water management. Here are some of the important agencies involved in this process:

- The Bureau of Land Management
- The Army Corps of Engineers
- The Environmental Protection Agency
- The Department of Agriculture.

There are other departments, such as the U.S. Department of the Interior, that get involved when more complicated negotiations are made with other countries.

State of California

In the State of California, there are many agencies who set policy specifically for California water management:

- The Department of Water Resources
- The California Environmental Protection Agency
- The Department of Consumer Affairs
- The California Department of Fish and Wildlife

Nonprofit Organizations

There are many nonprofits that influence policy within the state. While they may not be directly involved in managing water, they provide public influence and are active lobbyists in Sacramento. Some of these are well known. Others may surprise you:

- Heal the Bay
- The Environmental Defense Fund
- The Audubon Society
- The Sierra Club
- The National Resource Defense Council
- Surfrider Foundation
- Wishtoyo Foundation

Some of these organizations have gone to court to seek compliance with existing rules by government agencies.

Regional

Some cities and counties have teamed together to provide water management for their regions. Examples include:

- The Metropolitan Water District
- The Imperial Irrigation District
- Regional Water Quality Control Board

County

Counties and county regional organizations are also vitally involved in water management. Locally in Ventura County, there are several water management agencies and associations:

- Watershed Coalitions of Ventura County
- Association of Water Agencies
- County of Ventura Resource Management Agency
- Ojai Basin Groundwater Management Agency
- United Water Conservation District
- Fox Canyon Groundwater Management Agency
- Ventura County Stormwater Management Program
- Ventura County Public Works
- Ventura County Watershed Protection District
- Regional water suppliers such as the Calleguas, United, and Casitas Water Districts, and the Port Hueneme Water Agency

Responsibilities of more than one city or agency can be tied together by a document called a joint powers agreement.

Local

Then there are local water districts that directly serve a small population. Examples include:

- Camrosa Water District in Camarillo
- Many mutual water companies using their own wells
- Waterwork districts such as the Simi Water Works District
- Irrigation districts that serve only a smaller area by either purchasing water from a large purveyor or using their own wells

If the goal is to help you understand *how* to be involved in water policy in your area, knowing who manages your water and how they influence local water policy is important. By virtue of knowing this information, it makes it possible for individuals to learn and participate in decision making.

It's important to understand, particularly with federal agencies, that water policy is often handled by people who are appointed, not elected.

However, on the local level, such as with the Casitas Municipal Water District, board members are elected. So, if you're interested in how your water is managed, it's possible to contact your local water agency and get to know your elected officials.

If the saying is true that "all politics are local," then this applies also to local water politics. It's here that you can have an impact on your area's water management.

Because you may have chosen in the past not to be involved, many of your local water officials run uncontested term to term in elections. They are uncontested because the election is unknown to the general public. If after reading this, you decide that water management is important, then take the time to get involved. Know your water politicians. Perhaps you may even decide to run for office.

VII.

Extensive Effort, Minimal Impact: The Truth About Water Conservation

Governor Jerry Brown has been big on water conservation, especially with the announcement of mandatory water restrictions for municipal users in April 2015.

Here in California, if you read billboards or watch public ads on TV, the state government's mentality today is focused almost exclusively on conservation efforts. People using less water. Cities announcing mandatory water restrictions. Lawns going "gold."[29]

The problem with this approach is that it's simply not enough. In a recent Washington Examiner article analyzing

---

[29] For a recent analysis of California's misguided attempts to discourage and/or reduce the amount of urban greenery and lawns around the state, read Jeanette E. Warnert's "Drought concerns causing unnecessary impact on landscapes and lawns," University of California, Division of Agriculture and Natural Resources, 31 July 2015
http://ucanr.edu/?blogpost=18517&blogasset=74534

the state's focus on conservation, California's efforts were summed up like this: "The main focus of the SWRCB's (State Water Resources Control Board) conservation mandates represent just 10 percent of California's developed water (i.e., the water available for human use). Even if all Californians suddenly stopped using all municipal water, California water use would drop only 10 percent."[30]

Recent political battles in the state have involved big agriculture. As we have noted earlier, farmers are business people. When asked to conserve, they naturally calculate the cost of water versus growing crops. Many have simply chosen to grow less and leave land fallow. This doesn't solve the issue either because the problem is not conservation alone but to increase the actual amount of water we need to maintain our infrastructure and our existence.

California has driven itself into a corner by contributing to its own water management inefficiency. In short, there are too many chefs in the kitchen to make water regulation efficient, too many users making demands on a limited water supply, and too few water resources to supply the growing demand.

The result is a rising tide of panic among water agencies and politicians, and we as individual citizens can feel it. As noted by a recent drought bill introduced by

---

[30] Carson Bruno, "In California, water conservation isn't enough," *Washington Examiner*, 24 June 2015.

Senator Diane Feinstein, the official answer to drought management seems to be that we throw money at older technologies and hope that something sticks to the wall.[31]

Such an example is taking place right now in Santa Barbara, where years ago the city built a desalinization facility and by the time it was ready to use, the existing drought was over. The facility was mothballed. Now, to restart that same facility, it's going to be very expensive and environmentally controversial. The question remains: Is it worth the return on investment?[32]

One of the largest privately funded desalination facilities in the world is being proposed in San Diego County by a company called Poseidon. It's due to be completed and opened in 2016. But at what cost? The plant is expensive to build and maintain, and something else that might be troubling: Will a private company keep water costs manageable for the San Diego County Water Authority and its 3.1 million users?[33]

As we noted earlier in this book, California has too many people thirsty for water who live in a desert terrain

---

[31] Kevin Freking, "Feinstein's drought bill different," *Ventura County Star, Associated Press*, 30 July 2015.

[32] Amanda Covarrubias, "Santa Barbara working to reactivate mothballed desalination plant," *Los Angeles Times*, 3 March 2015.

[33] Matt Weiser, "Could desalination solve California's water problem?", *The Sacramento Bee*, 18 October 2014.

with limited water resources, and we're allowing more and more people to come to California. Even as they arrive, we're not telling them the honest truth about living in California: We don't have enough water.

The truth about water usage in California is that we've outstripped our resources and allowed too much growth for the amount of water that we're able to pump into businesses and homes, and even if current goals for water conservation are met, it's still not enough. We have managed to put ourselves in a bind that conservation by itself cannot solve.

What is actually required is a change in thinking. Our old model, which relies almost exclusively on water from snowpack in the mountains and from draining our rivers and lakes has proved faulty.

In addition, the mentality that we only address water needs during drought years is shortsighted. We should be planning and designing new technologies to increase our water supply, whether or not we have a long-term drought.

Conservation is fine as a public policy, but the impact on our overall water supply is minimal. The real discussion needs to become more focused on long-term planning, and it is imperative to look far into the future and think about replenishable and sustainable water supplies.

VIII.

Harvesting Rain and Snow:
The Unpredictable Nature of Weather

The rain and snow that are so familiar to us in California will eventually return, but the amount of rain we may receive in the future is unknown. However, one thing seems clear. As of 2015, the fourth consecutive year of drought in California is not just a random weather phenomenon. Something else is happening in our atmosphere and with our water resources. Something which we as humans have initiated.[34]

Even with predictions next fall and winter for a large El Niño, experts are saying that the precipitation we may receive would only be a drop in the bucket in relation to

---

[34] A succinct summary of this can be found in a recent *Rolling Stone* article by Eric Holthaus: "The Point of No Return," *Rolling Stone*, 13 August, 2015, National Affairs.

the amount of water we actually need to end the drought.[35]
With the impact of climate change, we may indeed be
tenuously waiting for the weather to solve our problems,
and that wait could last years or even decades.

The most important questions to ask regarding El
Niño's abundant rainfall are this: When it *does* rain, what
do we do with it, and how does it help solve our water
problem?

Let's suppose that we are lucky enough to have several
El Niños and that the drought is officially declared to be
over. Does this solve our problem? The answer is no. Even
if the aquifers were replenished, it still leaves us at risk for
future droughts because, as we have noted, our human
growth has outstripped what our natural resources,
particularly in drought years, can provide.

We have grown too dependent on water replenishment
from Mother Nature. As a result, human survival itself is at
risk. Complacency and ill-advised assumptions about
water replenishment have created a conundrum in which
we can no longer afford to simply sit and watch the clouds
in November and pray for rain.

There are two factors that create high risk for water in
California. One is that the normal annual snowpack we
have relied on for water has diminished. In April 2015,
Governor Brown traveled to a normal point of snowpack

---

[35] Ben Chou, "Will El Niño Save California from
the Drought?" *Switchboard: National Resources Defense
Council Staff Blog*, 6 August 2015.
<http://switchboard.nrdc.org/blogs/bchou/will_el_nino
_save_california.html>

measurement in the Sierras. Unfortunately, there was no snow to measure.[36] This could be due to prolonged drought, but there is another factor we might need to consider, and that is the impact of climate change.[37]

The direct effect of climate change on snowfall varies from place to place and year to year, but in California, it seems to be directly connected to our drought and its consequent lack of rainfall and snow.

Since we've historically been dependent on this weather cycle, the current drought only serves to emphasize that these cycles of rain and snow on which we heavily rely are inadequate for the number of people who need to be served. When this water cycle might change and improve is anyone's guess.

The question arises: Can we predict rainfall and snow in order to prepare for future droughts? And the answer is not yet. The further into the future we fly, the less accurate our abilities to know the future of our seasonal weather.[38]

---

[36] Zoë Schlanger, "Governor Brown Orders First-Ever California Water Cuts After Seeing No Snowpack,"*Newsweek*, 15 April, 2015, U.S.

[37] Catherine Gautier, "How climate change is making California's epic drought worse," *The Conversation*, 20 May, 2015.
<http://theconversation.com/how-climate-change-is-ma king-californias-epic-drought-worse-40030>

[38] For an explanation of the accuracy of current weather forecasting, read William Harris's article "Why can't scientists accurately predict the weather?" HowStuffWorks.com, 7 July 2010.
<http://science.howstuffworks.com/nature/climate-weat

We therefore conclude that we have been relying on some risky assumptions and unreliable data to supply water to citizens in California, and this approach needs to be changed in the long term if we are to survive this drought and others in the future.

But what can be done to deal with our impact on the weather and the environment? It's not enough to simply point out the problems. Our goal in this book is to find solutions, and this is what we are going to do. We don't want to recycle old patterns of adaptation. Our goal here is to point toward the future and suggest what new things might be done to increase our current water supply.

Thus, after extensive review of how we've reached this point of crisis, we've come to the high point in our book. We're now going to look into the future. We contend that there *are* solutions. For us, for governments, for water policy makers, for water buyers and sellers, for farmers, for cities, for everyone who is involved in using water, it's time for a change in our habits. Conservation? Yes, but also new technologies, new solutions. That is the key to our own survival.

---

her/atmospheric/scientists-predict-weather.htm>

IX.

The Future is Now:
Solutions to Increase Our Water Supply

The good news is that there are many, many ways to increase our water supply. Beyond reverse osmosis and desalination. Beyond water conservation. Beyond the aqueducts that tend to drain our rivers and lakes.

We are going to suggest some alternatives that, while they may not be new ideas, have certainly been underutilized in planning and increasing water resources. Here are a few suggestions that will actually, even in this drought period, help us have *more* water as a near-term and long-term solution.

Water Recycling

Water after it's used is known as wastewater. Wastewater goes through four process before it is released for reuse. The first is called **primary**, which allows solids to settle and the water from the top to be used. Then it could go for another treatment. **Secondary process water** is treated with a filter bed, using such things as rocks, known as biofilters. Some places add oxygen in this process. **Tertiary treatment** involves adding different chemicals like chlorine for disinfection, or removing things like nitrates that could be harmful if discharged into natural streams.

Beyond the **primary, secondary,** and **tertiary** water filtering process, some advanced technologies and ideas are already in use. One such treatment is the reverse osmosis or desalination which falls under **advanced water treatment**, examples of which are the advanced purification facility in the city of Oxnard and the brackish[39] water reclamation facility used by the Port Hueneme Water Agency. In Oxnard, they have the capability of taking water from their wastewater treatment plants and recycling it for such uses as agriculture and greenery. Port Hueneme takes water directly from the United Water District, which comes from underground water holding zones called aquifers.

Both of these plants result in purified water that is so good, it would corrode the pipes if not blended with other kinds of water. Recycled water[40] uses a significant amount

---

[39] Brackish water has more salt than drinkable water. Usually human consumption is about 600 parts per million (TDS) water. If the number goes to 1,000 parts per million (TDS) it is not drinkable.

[40] Recycled water is wastewater or groundwater or any other type of water which is used for many different purposes, such as median irrigation, agriculture, and keeping ocean water back that has intruded into freshwater aquifers.

of water which would otherwise be wasted.

If you're interested in the real-time process of water purification, most cities such as Oxnard and other local water agencies will be glad to give you tours and explain the mechanics of their work at no cost to you.

Water Capture

Capturing, simply stated, is the use of water that we don't capture today, such as rain water, storm water, and flood water, and using it for other purposes such as irrigation. It's known as "harvesting" rainwater.[41]

There are many places in the world where they capture rainwater. Generally it goes to an underground cistern and is pumped from this cistern throughout the year. Common terms for stormwater reclamation include "integrated water management" and "low impact development."[42]

This is an idea that is catching on. In Los Angeles, for instance, the Department of Water and Power recently

---

[41] The Los Angeles Department of Water and Power has a new plan in place to capture stormwater. For more information on this, read Monte Morin's article, "Why Waste Rain? DWP has plan to capture stormwater that now enters the ocean," *Los Angeles Times,* 25 June, 2015, California.

[42] For more in-depth explanations of stormwater reuse, refer to Wikipedia's "Stormwater." <https://en.wikipedia.org/wiki/Stormwater#Stormwater_management>

announced a plan to capture storm water.[43] The County of Ventura has implemented a similar type of storm-runoff project for farmers and residents in the upper Ojai Valley.[44]

Flood waters can be captured using temporary dams in a river. These inflatable dams are made with geosynthetic fabric and can be installed right before a large storm comes in. They divert water and store it in a depression, aquifer, or abandoned quarry until the water can be reused. [45]

Salton Sea as a Reservoir

One of the most impactful projects being proposed is at The Salton Sea, a large body of water located northeast of San Diego in Riverside County.

---

[43] Monte Morin, "Why Waste Rain?" *Los Angeles Times,* 25 June 2015, California.

[44] Tim Pompey, "County Agencies Cut Ribbon for Groundwater Recharge Project," *Tri County Sentry,* 25 September, 2014 <http://tricountysentry.com/blog/county-agencies-cut-ri bbon-for-groundwater-recharge-project>

[45] Susan Kline, "Building the better mousetrap for rainwater," Citizens Journal, 29 April, 2014. <http://citizensjournal.us/building-the-better-mousetrap -for-rainwater>

Technical and engineering information for the temporary inflatable dam by Mohammed Hasan and Mark Capron is available at: <http://oceanforesters.org/uploads/Water_Crop__Farm _Bureau____BuRec__Mar17-11.pdf>

It's unique in that it's the only inland sea that we have in California. It was formed centuries ago because the elevation of the Salton Sea is below sea level. Thus, when the Colorado River flooded, it created the Salton Sea. It is actually a body of salt water, even saltier than ocean water.

The Salton Sea is an incredible water resource that is currently dying off. If nothing is done to preserve it, this sea has a million acre feet of water evaporating, and if it keeps evaporating, it will dry out and create an environmental hazard because the wind will blow toxic dust that can literally spread all over Southern California. Some irrigation districts dump regular water from the Metropolitan Water District into it just to keep it alive.

The Salton Sea problem can be dealt with and/or eliminated in two ways:

- If we build an aqueduct from the Gulf of California to bring gulf water into the sea.
- If we use the groundwater that is highly salted, purify it, and dump it into the sea.

The reason these methods may be useful is that the Salton Sea itself can be used as an additional water reservoir, which can be purified using new technologies such as geothermal and solar desalination as proposed involving Waterfx.[46]

Some may ask why we don't build a regular reverse osmosis desalinization plant like the one in Santa Barbara

---

[46] <http://waterfx.co>

at this site. The problem with reverse osmosis in its current stage is that it creates a brine, a waste product that has to be dumped somewhere else. It is not recommended to dump this without further treatment in the ocean because it may be environmentally toxic to fish and other wildlife. The advantage of the desalination proposal involving Waterfx is that its byproduct is solid salt, which is reusable.

Other methods on the cutting edge of water purification include:

* Capacitive deionization technology systems[47]
* Flash distillation[48]

For Hasan Consultants, in cooperation with Ocean Foresters[49], our basic proposal for recapturing usable water from the Salton Sea is a highly technical engineering process. In simple terms, it goes something like this:

* Build an aqueduct from the Gulf of California (also known as the Sea of Cortez) to the Salton Sea to replenish its supply of water; or
* Dump groundwater from the aquifer below the Salton

---

[47]

<http://www.sciencedirect.com/science/article/pii/S0019916408002294>

[48]

<http://www.rpi.edu/dept/chem-eng/Biotech-Environ/Environmental/desal/flash.html>

[49] <http://oceanforesters.org>

Sea that holds many thousands of million acre feet of water back into the sea and then treat it.

The plan would help solve and offset the potentially expensive environmental costs that await Californians if the Salton Sea dies. It would also take pressure off of the Colorado River and allow it to be replenished. It's even possible that it might provide more water for people along its shore as well as let it reflow naturally into the Baja Gulf, revitalizing former estuaries that used to be a part of its natural path.

It's not an easy engineering design, nor is it cheap, but when the Salton Sea is replenished and reclaimed by the aforementioned technologies, it has tremendous potential to provide water that could be pumped all across the Southwest to areas like California, Arizona, Nevada, and Mexico. It is estimated that the revived Salton Sea could provide a new water source for as many as 44 million people.[50]

---

[50] For a technical engineering plan that we call the "International Integrated Water Project," read an article in the online news site Citizens Journal:
<http://citizensjournal.us/events/event/international-integrated-water-project-iiwp-startling-innovation-in-water-supplies>

Engineereing and technical information is available at:
<http://oceanforesters.org/uploads/IIWP__California__May1__2015.pdf>

X.

A Call to Act:
You *Can* Be Involved

The danger we face is this. If something is not done now, it will take generations of suffering to catch up to our current problems. The good news is that all of this can be prevented by sustainable projects and technologies that we already have. Since it will take a lot of money and a lot of time and a lot of political willpower to build such large projects, it should be started now in order for us to reap the benefits. Otherwise, waiting just makes it that much harder to solve.

It's important in this debate about the future of water in California to inform and empower everyone who has an interest in this issue about how they can help change and improve the process.

- Through personal research and education
- By engaging in discussions with like-minded people

- By studying the various technologies that up to this point have been dominated by the voices of innovative scientists and engineers
- By becoming socially and politically active with water issues within your city, your county, and your local water district
- By visiting the website of purveyors to understand their individual way of managing water. Web sites also have reports and documents that would be very interesting for citizens to read.
- Attend board and council meetings when they are discussing water issues.
- Ask questions when you need clarification.

Our goal in this book has been to spark thought and encourage discussion. We will not agree on all the issues, but there is something about human discussion that sorts through and finds a reasonable path to follow. That's the way societies seek to survive and make progress.

If at some point in hearing about water, you have been overwhelmed by the size of the problem, we hope this book can unravel some of these thorny issues and at least clarify what's happening in our state and around the world.

Individual citizens do not have to feel powerless. They are asked to sacrifice so much, but once they understand and comprehend the needs, they also can participate in the solutions.

This simple book allows every citizen to feel knowledgeable enough to participate in these solutions when it comes to dealing with drought and water supply. From crisis to action, everyone, both citizen and scientist, can participate in solving California's water shortage problem. What matters most is that we work together and think far into the future. Ask the question: What might be accomplished if we pool our thoughts and resources? What, indeed? That is the exciting thing about this book. We know not where it takes us, but we know simply by learning that the answers are out there. The water crisis is solvable. It's not impossible or even improbable that we can find a better way to have water for everyone. It just requires us to think cooperatively, and, most important, to believe in ourselves. The water is there. We encourage you as a reader, whether a scientist, engineer, or citizen, to go out there, find new water, and help make it available to everyone here in California.

www.ingramcontent.com/pod-product-compliance
Lightning Source LLC
Chambersburg PA
CBHW071004180526
45168CB00003B/1285

* 9 7 8 1 5 1 6 9 0 6 7 9 6 *